THE HARMON MEMORIAL LECTURES IN MILITARY HISTORY

NUMBER THIRTY-TWO

United States Air Force Academy
Colorado
1989

AIR POWER, ARMIES, AND THE WAR IN THE WEST, 1940

R. J. Overy
KING'S COLLEGE
UNIVERSITY OF LONDON

For a complete listing of previous Harmon lectures see pages 22, 23, and 24.

Views or opinions expressed or implied in this publication are those of the author and are not to be construed as carrying official sanction of the Department of the Air Force or of the United States Air Force Academy

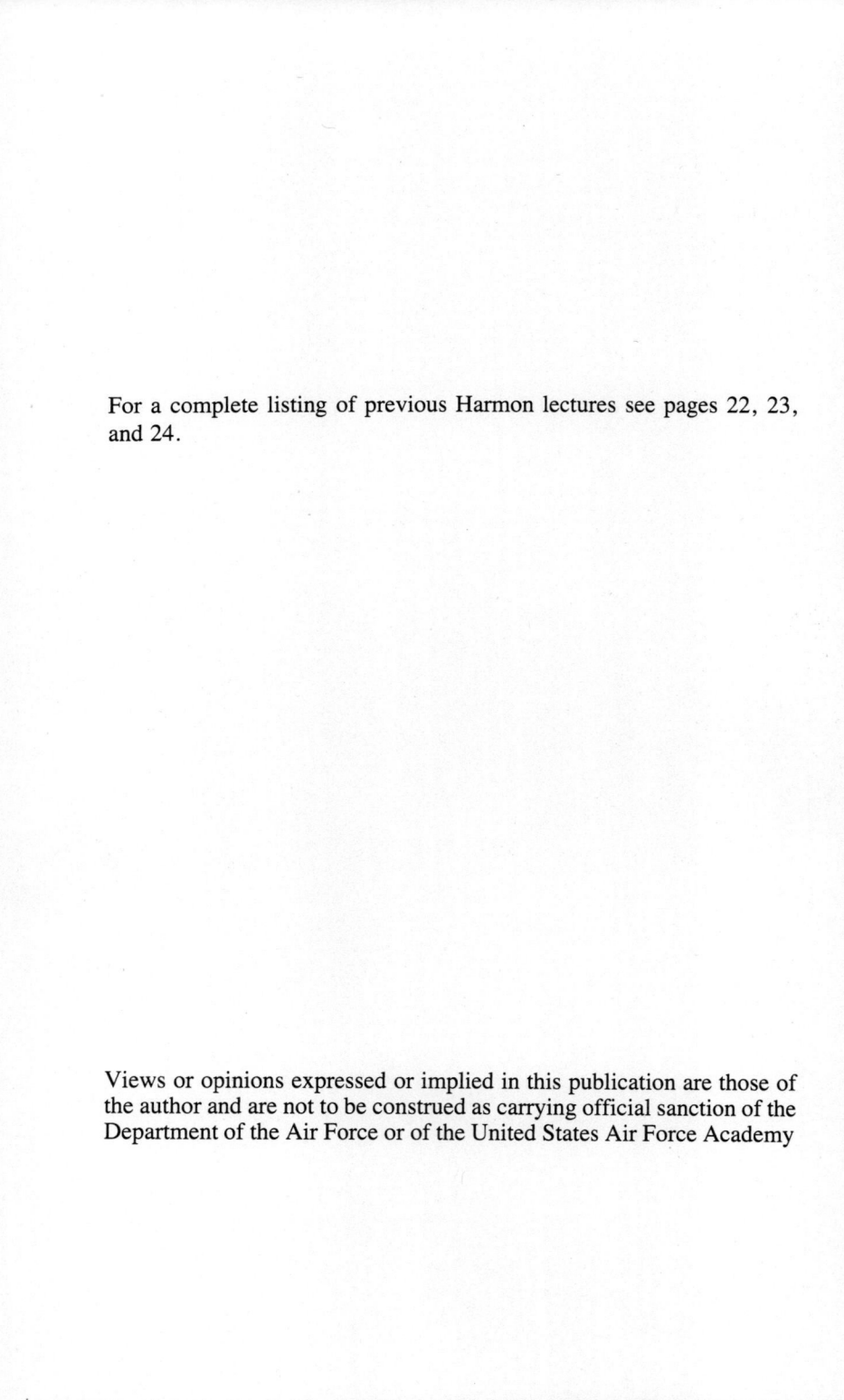

AIR POWER, ARMIES,
AND THE WAR IN THE WEST, 1940

R. J. Overy
King's College, University of London

THE HARMON MEMORIAL LECTURES
IN MILITARY HISTORY

NUMBER THIRTY-TWO

United States Air Force Academy
Colorado
1989

For sale by the Superintendent of Documents, U.S. Government
Printing Office, Washington, D.C. 20402

THE HARMON MEMORIAL LECTURES IN MILITARY HISTORY

COMMITTEE

CHAIRMAN: Colonel Carl W. Reddel, USAF
Professor of History
U.S. Air Force Academy

MEMBERS: Harold C. Deutsch
Retired

John F. Guilmartin, Jr.
Associate Professor of History
Ohio State University

Richard H. Kohn
Chief, Office of Air Force History

EXECUTIVE SECRETARY: Captain Karl H. Zimmerman, USAF
Instructor of History
U.S. Air Force Academy

**Lieutenant General
Hubert Reilly Harmon**

FOREWORD

The Department of History, United States Air Force Academy, is the sponsor of this lecture series. The purpose is twofold: to promote a greater knowledge of military history in the United States and abroad, and to stimulate a lifelong interest among cadets in their professional subject.

Each year a committee of internationally known civilian historians and Academy representatives selects an outstanding military historian, or a first-rate scholar from a closely related field, who is invited to present an original lecture on a subject of his choice within the broadly construed field of military history. In keeping with the purpose of the series, the lecture is published and distributed to leading scholars and libraries throughout the world.

These lectures are known collectively as the "Harmon Memorial Lectures in Military History," in memory of the accomplishments of the late Lieutenant General Hubert R. Harmon, the first Superintendent of the Academy. General Harmon's lifelong personal interest in military history makes it particularly appropriate that he be honored in this way.

R. J. Overy presented the thirty-second lecture in this series on Tuesday evening, 28 November 1989, to members of the Cadet Wing and faculty, as well as distinguished guests. "Air Power, Armies, and the War in the West, 1940" was extremely well received and offered new insights on the use of air power in the early years of World War II.

Dr. Overy is an internationally known historian of World War II. Born on 23 December 1947, he earned his B.A., M.A., and Ph.D. from Caius College, Cambridge. He served as a Research Fellow at Churchill College from 1972-73 and as Fellow and Assistant Lecturer in History at Queen's College from 1973-79. Since 1980, he has lectured in history at King's College, University of London.

He has also written extensively on air power and the Third Reich as well as the British Empire. His books include, *The Air War, 1939-45* (1980), *Göring: The "Iron Man"* (1984), and, most recently, *The Road to War* (1988), a study of the origins of World War II which has also been the basis for a major television series. He received the T. S. Ashton Prize from the Economic History Society in 1983 for his article on "Hitler's War and the German Economy," and the Cass Prize for Business History in 1987. He is currently writing a history of the Nazi economy and a major study of World War II.

AIR POWER, ARMIES, AND THE WAR IN THE WEST, 1940

Harmon Memorial Lecture U.S. Air Force Academy

28 November 1989

R. J. Overy

It is now almost fifty years since German armies routed British and French forces on the northern plains of France. It was a victory almost unique in twentieth-century warfare in its speed and decisiveness. So rapid and well-planned was the German advance that it was not difficult to argue that this was just the campaign for which Hitler had long been preparing. German victory strengthened the view that the Western Allies stood before him weakened by years of military neglect and political feebleness. Since the war, this view has become embedded in popular wisdom. There is a strong consensus that the Western powers were militarily unprepared, much weaker than the enemy they faced, and that German success rested on exploitation of the new air and tank weapons. This new technology was integrated in the Blitzkrieg formula, which cruelly exposed the poor planning, strategic bankruptcy, and low morale of Germany's enemies. The core of German success, so the argument goes, was the overwhelming air power that Germany brought to bear on the conflict. At least one historian of the campaign has concluded that it was "the Luftwaffe's supremacy in the air which constituted a decisive factor."[1]

Though there would be little point in denying that the air force did constitute an important element in Germany's success in May and June 1940, there remain nonetheless a great many questions that historians can still ask about the role of air power in the campaign. We must dispense, first of all, with some powerful misconceptions about the nature of the conflict. The first and most enduring of these is the belief that there existed a great disparity of air strength between the two sides. The facts show otherwise. Germany massed 2,741 combat aircraft for the campaign in the West on 10 May 1940, 1,000 of them bombers, 970 of them single-seat fighters.[2] On the same day, France possessed over 3,000 modern combat aircraft; 2,500 modern fighters had been supplied to the French Air Force during 1939 and the first five months of 1940. The difference was that French forces were very dispersed: only 500 fighters were stationed in northeastern France opposite German forces, and 1,000 aircraft were in North Africa.[3] The Royal Air Force (RAF) had stationed about 250 aircraft in France in May 1940, half of them bombers, half fighters—but the RAF as a whole possessed by then over 2,000 modern first-line aircraft, including 700 advanced single-seat fighters in the home squadrons.[4] It is striking that Britain initially devoted to the French campaign not many more aircraft than Belgium, which had 184 aircraft operational in May 1940.[5]

Taken together, the French and British air forces could have mustered over 4,000 modern, combat-ready aircraft, and almost a third more fighter aircraft than the Luftwaffe. These were much better odds than the RAF faced a few months later in the Battle of Britain. Moreover, both Western air forces had a higher level of reserves (mainly of technically inferior aircraft) than the Germans and by May 1940, with American aircraft supplies, had a monthly output of aircraft more than double German levels. By the end of the Battle of France, it has been calculated that the French Air Force alone had been supplied with 2,900 modern fighters since 1939. Indeed, it had as many combat-ready aircraft in fighter squadrons at the end of the battle as it possessed at the start.[6] It is certainly possible to argue that the German Air Force had a marginal lead in quality, particularly in bombing aircraft rather than fighters. But in terms of numbers of aircraft the gap disappears. The issue this raises is an important one: why did the Allies fail to use their numerical and production advantages in an effective, concentrated way?

The second misconception stems from the view that the Western states failed to understand the new weapons of war, and displayed a particular incompetence in the exercise of air power. There were, of course, plenty of mistakes and false hopes. It is not difficult to construct a largely negative view of Western achievements in the air. But this perspective neglects two essential points. First, the British were only weeks away from winning perhaps the only major air force-to-air force conflict of the whole war, the Battle of Britain, and by a less narrow margin than the British like to think. Second, both British and French military thinkers and planners had worked away at the problems of air power with every bit as much energy as their German counterparts. Western planners had formulated a role for air forces and integrated them into the general strategic plan. It could certainly be argued that some of the choices they made turned out to be inappropriate for the campaign they fought. But defeat in 1940 occurred not for want of thinking hard about the use of aircraft. The issue again is to ask why the Allies made the choices they did about air strategy; their choices were not made from mere incompetence or perversity.

A final misconception lies in the view, widely held, that the odds were always stacked against the West from the start, and that their defense of Poland was a vain, belated, and doomed moral gesture in the face of certain defeat. Yet this view, too, ignores that neither state declared war expecting to lose. They planned in detail the war they would fight. The war plans drawn up in the spring of 1939 worked on the sensible assumption that war could only be contemplated if the chances of victory were considerable. British and French strategy rested on the view that Germany could be contained in 1940 by a combination of naval blockade—and it should be recalled how much more substantial Western naval strength was compared with German—together with the prepared "continuous front" from the Alps to the North Sea, and bombing from the air against Germany's "vital centers." The British draft of the War Plan in March 1939 argued that with Germany isolated diplomatically, hemmed in by the Maginot system, and with the economic might of the two Western empires mobilized to the full "we should regard the outcome of the war with confidence."[7] Anglo-French military planning was not some grand suicide pact, but a genuine reflection of growing belief in Western dispositions and staying power.

With this said, however, it is still possible to argue that in the campaign fought in May 1940, German forces proved greatly superior in fighting power, organization, tactical flexibility, and coordination of forces. This observation was particularly true of the German Army, and it would be wrong to understate the role played by the German Army, as distinct from the Air Force, in the campaign.[8] Modern air power was untested except against the lightest resistance in Spain or Poland. The German planners placed their hopes in the strategic novelty of the army's so-called "sickle-cut" strategy based on a concerted armored drive through the Ardennes. It is striking how little air power featured in the strategic and tactical discussions of the planned attack on France. When the air force was used, it conformed closely to German Army doctrine and the army's high operational standards. Broadly speaking, this doctrine can be reduced to three "C's"—concentration of force, counterforce strategy, and coordination of forces. There was nothing very original in this approach; German planners looked at the core of the problem, which was to attack and decisively destroy the military forces of the enemy. They hoped to do so with optimum operational effectiveness, economy of force, and above all, concentrated and combined effort. But they were far from confident in 1940. The armed forces had not expected a general war to break out in 1939; neither had Hitler. Alternatives were explored: sitting defensively behind the Siegfried Line; or, the brainchild of the younger Luftwaffe staff officers, an all-out air assault on Britain in the late autumn of 1939 to knock her out of the war.[9] But in the end the offensive tradition in the German forces, combined with a recognition of geopolitical reality, pushed them, and Hitler, to the view that the decisive land campaign was the only real option. The Battle of France was thus no well-prepared, single-minded campaign. No Schlieffen Plan existed in Germany in 1939. The campaign was the hasty product of staff planning over the winter and spring, recognizing German limitations and making the most of German strengths. It was the Allies, not the Germans, who had a clear idea before September 1939 of the kind of war they were going to fight.

We can now move beyond the arguments that it was unequal strength and lack of preparation that doomed the West from the start, to look at the question of why the three states involved in the battle organized, prepared, and deployed their air forces in such different ways and with such contrasting effect. The German case is by now well-known, but it is worth recapitulation. The preparation for war was dominated by the army view of strategy: the role of the armed forces was to constitute an offensive force, capable of attacking and defeating the enemy armed forces in the field. This attitude made geopolitical sense considering Germany's long exposed borders, and the recognition that any future major war was likely to engage the two largest land armies, French and Russian. These views also complemented the preconceptions of the Commander in Chief, Hitler, whose own ambitions dictated that Germany should wage offensive war.

The general instruction for the Luftwaffe, the "Conduct of Air Warfare," first drawn up in 1935 and revised down to 1940, conformed to this general military outlook. The Luftwaffe's goal was "to defeat the enemy armed forces" and to do so in combination with the other services.[10] The operational guidelines were straightforward. The

Luftwaffe was to be used first as a counterforce instrument to destroy enemy air power and establish air superiority over the campaign area; then it had to concentrate on the task of giving direct support to the land campaign. This support was to consist of providing a protective umbrella over German armies, attacking targets near the front-line, and attacking rear areas where new troops, communications, or supplies were located. Only in the event of stalemate and the expectation of a long, drawn-out war would the Luftwaffe bomb enemy vital centers. The lessons of Spain and China and, of course, Poland, suggested that direct tactical support would be the most effective use of air power. Even Colonel Walther Wever, the Luftwaffe Chief of Staff who died in 1936 and who is generally regarded as the chief German proponent of strategic air warfare, argued that "in the war of the future, the destruction of the armed forces will be of primary importance."[11] The Luftwaffe staff was filled with ex-soldiers or ex-World War I fighter pilots who broadly shared this view. Combined service maneuvers from 1935 onwards demonstrated conclusively that tactical air power would bring the best results given the current technology.

German air planners closely linked technology to strategy. Given the limitation of aircraft in the late 1930s, they concentrated on fast fighters, medium bombers, and dive-bombing aircraft which could perform the combined operation function. Great emphasis was placed on front-line communications. In 1940 the Luftwaffe possessed three signal regiments, 63 signal companies, and 115 special signal units to coordinate air attacks and air-ground cooperation.[12] Yet the Luftwaffe did not ignore strategic bombardment, as is so often claimed. In 1938 the long-range Heinkel He-177 bomber was put into the production plans; a year later an intercontinental bomber (the so-called "Amerikabomber") was also included. But these weapons would not be available in quantity until 1942-3. Luftwaffe operational surveys from autumn 1938 to late 1939 demonstrated that with the current technology nothing decisive could be achieved from the air against Britain until that date. Although Göring wanted a more vigorous strategic role for air power, and the younger staff officers were eager to send Britain the knock-out blow she was so manifestly expecting, the German high command kept the air role within its technical limits, and did so to very great effect.[13]

There is little here that is surprising. More interesting is the fact that Britain and France, both states that had waged, like Germany, a largely tactical air war on the Western Front in World War I, failed to do what the Luftwaffe did. Here the French example is the more remarkable, for France shared much of the military and strategic outlook of her erstwhile enemy. Like Germany, France was governed by geopolitical necessity and the search for resource efficiency in warfare. Like German forces, the services were dominated by the army and by army interests. The organization of the air forces in the two states was not strikingly different in the late 1930s except in one regard: the separate army units in France expected to have air units assigned for their individual use, while in Germany, air units were assigned in "fleets" to support whole army corps. This distinction was to prove a very significant difference in approach to tactical warfare. At the level of strategy, the chief contrast was that French leaders geared their forces to defend France, not to promote an offensive. The whole of France's large and expensive military effort in the interwar years was based on the establishment of a "continuous front" with an elaborate, prepared battlefield on which the German attack would be blunted. Once Germany was contained, the French would wear the enemy down by bombing and blockade until an offensive could be launched with any prospect of success.[14]

Given that the French strategic profile was different from the German—a defense strong enough to absorb and deflect the offense—the French view of how to use aircraft was remarkably similar. The French General Staff placed emphasis on the defeat of the enemy forces and saw air power contributing to this strategic aim. The *Instruction* of 1936 for the air force gave it three functions: providing a defensive umbrella over the mobilizing armies as they entered and established the prepared battlefield along the fortified front; air attacks on the advancing enemy, on concentrations of his troops, supply columns, strong points, etc.; and finally, attacks by bomber aircraft against rear-area targets. These bombing attacks were supposed to complement the battle-area attacks and were to be directed at an area no more than 200 kilometers from the front line, where only militarily useful targets should be attacked, such as communications and ammunition dumps.[15] Only in exceptional circumstances would attacks against industrial or city targets be endorsed. French military leaders were on the whole unimpressed with the distinction between tactical and strategic air power and opposed attacks against civilians. French air theorists regarded battlefield support as "strategic" air power inasmuch as it contributed to the general strategic aim. There were arguments among French military leaders about how to organize and command air forces, but by the late 1930s there was general agreement on the need for large tactical air forces to establish air superiority over the front and prevent the breakthrough that had destroyed French arms in 1870 and had almost done so again in 1914.

French air strategy, then, differed little from German. But in operational preparation and technical development there were marked differences. The French Air Force failed to develop either battlefield assault planes or a dive-bomber. Too much emphasis was placed on reconnaissance, artillery spotting, and army cooperation aircraft in the 1930s, reflecting the priorities of the ground army. When rearmament accelerated in 1938, the industry rushed out large numbers of good quality fighters and medium bombers. Not until 1940 did the French order large numbers of dive-bombers from the United States when, belatedly, they realized that they lacked any effective battlefield support planes.[16] Even more bizarre for a military that placed so much emphasis on the prepared battlefield and static, centrally controlled operations, the French forces had very poor communications. This deficiency was true of links between air units and between the air force and army. Only 0.15 percent of the military budget between 1923 and 1938 was spent on communications.[17] Radar was hastily imported from Britain in 1939 but hardly featured in the 1940 campaign. Poor communication fatally weakened French combined operations.

Another operational problem was the failure to assign an order of priority to the targets designated for the air force. The French Air Force did not develop a counterforce strategy as such. Attacks on the enemy army were regarded as being of equal importance. Second, the air force failed to impose on the army an organizational plan that would permit the concentration of effort that the *Instruction* to the air force required. Air power was parcelled out, like tanks, to support each area of France where army units were stationed, and to support each division at the front. This organizational structure was the end-product of a long argument between the air force and army which spilled over into French prewar political conflicts. In 1936 the new Popular Front Air Minister, Pierre Cot, favored counterforce and concentration of effort as well as a more independent air force to make this possible. The army opposed Cot because he was left-wing and challenged the army's military monopoly; the fall of the Popular Front in 1938 permitted the army to dominate air strategy again and to decentralize air forces to meet army requirements. What the French Air Force lacked was a Trenchard or a Göring who could fight their political battles for them while they got on with the job of preparing for air warfare. Instead the French got the worst of both worlds: a sulky air force, eager to organize its own campaign, and a jealous army equally motivated to use the air weapon in its own way.

The British approach to air strategy differed in almost every respect from the continental powers. This uniqueness was due in no small part to the RAF winning its independence at the end of World War I, allowing it to pursue what it self-consciously saw as air strategy in its own right. The reality of Britain's geographical position and her imperial obligations, and the political weakness of the British Army, also influenced British air power development. But the fundamental difference lay in the kind of war the British expected to fight and in their strategic priorities. General Golovine's "Air Strategy," published in London in 1936, illustrates this point exactly. Golovine wrote that British air strategy could be reduced to three problems: defending the British Isles, defending the Dominions and colonies overseas, and protecting the imperial trade routes. The RAF's contribution to this strategy was to secure the local defense of Britain and the Empire and to perform "general strategic duties" in the form of an "active defence." Active defense meant, in fact, offense, the bombardment of the enemy state as a means of weakening its military power and shortening the war.[18] Though Golovine's views were not official doctrine, they closely reflected the views of leading airmen. Air power was generally regarded as a means of prosecuting a long war of attrition in the style of a naval blockade from the air. The RAF insisted that it was not like the other services whose task was to seek out and destroy the enemy army or navy. The enemy air force was regarded as an indirect target. According to the *Manual of Combined Operations* issued in 1938 and still current in 1940: "Air strategy consists of attacks aimed at the destruction of . . . one or more of the enemy's vital resources." The best means of defense was therefore offense against enemy sources of industrial supply. This "indirect" exercise of air power dominated RAF thinking throughout the prewar period. [19]

Such a strategic priority affected the development of British air power in some obvious ways. The RAF consistently argued against a direct counterforce strategy. Air leaders regarded an attack against an air force as an "uneconomical expenditure" of effort. They believed that an enemy air force would always be too well-dispersed and too well-defended to be a profitable target. Tests conducted in the late 1930s against "airfield" targets confirmed this view. Not only were dispersed aircraft difficult to hit, but damage to runways could be repaired "in hours." When operational plans were produced in August 1939 for a campaign against the German Air Force it was found that only 12 percent of its airfields could be reached from British bases, and only 60 percent from French.[20] Airfields, it was argued, would have to be bombed at night, which raised navigational and bombing accuracy problems. The only effective solution to reducing enemy air power, according to RAF planners, was to mount "attacks against the aircraft industry." This view prevailed throughout the campaign in 1940, even when the British Chief of Staff directed the RAF in June to attack airfields and German aircraft on the ground.[21]

To the rejection of conventional counterforce strategy must be added the almost complete rejection in the RAF of a direct tactical role in support of the army. The *Manual of Combined Operations* devoted only three and one-half pages out of 272 to air-army cooperation and one-third of that was devoted to "Control of Semi-Civilised Tribes within our own Jurisdiction."[22] Of course it was true that until the mid-1930s it was not clear that a major land army would ever be raised and launched again on the Continent, and strategic air attack was seen as an alternative to trench warfare. But when it became clear in Spain and China that air-army cooperation had played an important role, and once Germany had become the main potential enemy, there was every case to be made for developing a tactical capability. Some British politicians began to suggest just such a course. Yet the RAF continued to urge strongly that neither its technology nor strategic profile were suitable for direct support of a land army. Further bombardment tests showed that railways were difficult and expensive to attack and fleeting targets of opportunity behind a moving front were virtually worthless given the time it would take to bring heavy bombers into action.[23]

The RAF view was that the target had to be large, preferably static, and important to the enemy war effort in a more permanent sense to justify the expenditure of operational effort and bombs in attacking it. If air power involved concentration of force for the British, it was concentration against well-defined, if complex, industrial target systems, not against airfields, supply lines, and troops on the move which were all difficult to hit, likely to be well-defended by flak and fighters, and incapable of decisive destruction. The British saw air power as an instrument for the big strategic gesture, not the minor tactical target. Battlefield targets, it was suggested, could more usefully be attacked by ground artillery. Even after the failure in France in 1940, the RAF inquiry into the campaign stressed that the failure in the air was due to the French insistence on tactical support which was "unprofitable" and directed at "random objectives."[24]

This view prevented the RAF from making any serious technical or organizational preparations to meet the needs of a tactical air campaign. The dive-bomber was examined, tested without enthusiasm, and rejected. High-level bombing was shown by bombing tests to be just as effective and less dangerous for the pilot. This conclusion was largely special pleading. Dive-bombing tests had in fact been carried out with aircraft that could barely dive, but merely glided downwards towards their target at an angle of 17 degrees. The margins of error recorded from these limited tests—averaging 82 yards from 3,000 feet—were felt to be too great to be worth using against battlefield targets.[25] When the Air Staff finally ordered more dive-bombing tests in spring 1940 with the one aircraft—the slow and vulnerable Fairey Battle—which could dive more than 40 degrees, the Advanced Air Striking Force in France trained a mere seven pilots, who dropped 56 bombs in practice.[26]

The RAF attitude to tactical aviation was exemplified by the specification for a new army cooperation aircraft laid down as late as March 1939. The specification called for a slow two-seater aircraft which could undertake "close and distant tactical reconnaissance by day, observation of artillery fire, photography, low-level or shallow dive bombing, and supply-dropping."[27] The call for a battlefield jack-of-all-trades came at just the time that greater functional specialization in aviation was evident everywhere, while the technical performance required made the aircraft obsolescent even at the planning stage. John Slessor, Director of Plans in the Air Ministry, expressed the widely-held view that "the aeroplane is not a battlefield weapon."[28] Small wonder that almost no preparation was made to provide effective battlefield technology or to establish systems of communication to cooperate with the army that matched the excellent communications system set up by Fighter Command in Britain. Small wonder, too, that the RAF fought all the way along the line not to be subordinated to the French Army in 1940, and not to divert more than a tenth of British air strength to the Battle of France.

There were, then, reasonable grounds for British arguments against tactical aviation. But they made Britain a less-than-effective ally for the French when they were finally faced with the day of reckoning. Under pressure from the French General Staff, the British military chiefs accepted a continental strategy in the spring of 1939. This meant a continental expeditionary force and an air force to serve in France. But when war broke out, both sides were still arguing about air strategy. Bomber Command wanted immediate attacks on Germany in September 1939 because she was "politically rotten, weak in financial and economic resources, and already heavily engaged on another front."[29] Gamelin, the French Commander in Chief, insisted that the air forces should be held back to attack the German offensive when it came.[30] Under political pressure the two sides finally agreed that a British air contribution would consist of a small striking force of bombers to attack the advancing enemy, and a separate air component under the French Commander in Chief for attacks in direct support of the army and the protection of Allied airfields. Slessor reminded his seniors of the strong opposition of his staff to the idea "that we are going to direct every aircraft against German columns on the roads."[31]

In practice, the British Chiefs of Staff made little effort to give clear priorities to the RAF, which continued to resist the tactical role throughout the period to June 1940, in anticipation of the assault on more valuable industrial targets in the Ruhr. This ambivalence explains the poor communication between the Allies, the lack of cooperative exercises, and the very negative research findings the RAF sent to the French on tactical aviation. Asked what Bomber Command could do to hold up the German advance, the RAF replied that tasking its medium bombers at full capacity for a week would only keep three German railway lines out of action.[32] But the French aggravated the problem. Gamelin insisted that the greatest caution be exercised in bombing missions to avoid initiating what the British called the "gloves off" conflict, attacks against civilian industrial targets. Rather than bomb railway stations and marshalling yards, the French Air Force Commander asked the RAF only to bomb railroads at some distance from habitation.[33] The RAF was not only asked to perform a role for which it had not been properly prepared and against which it was strongly prejudiced, but to do so with what it regarded as serious operational constraints.

There were other important considerations that governed the attitude of the Allies regarding air power. Their intelligence indicated (wrongly, as it turned out) that combined German and Italian forces vastly outnumbered their own and were backed with reserves of equal numbers, a total of over 12,000 aircraft.[34] This information encouraged the sensible view that the enemy should not be met head on since it would mean the immediate destruction of Western air forces; it discouraged the politicians from taking the "gloves off" for fear of massive retaliation on their own civilian populations. This intelligence also encouraged the British to keep the great bulk of their forces in the British Isles, and the French to disperse their forces over the whole of France, guarding the industrial regions and anticipating Italian intervention. The British refusal to transfer their air power to French soil was a critical decision. It was very difficult for the RAF to accept that Germany might not use the Luftwaffe to attack British cities, having argued for so long that it would, and British politicians were not prepared to take the risk of denuding the home country of air forces. Yet this outcome reduced any prospect for serious concentration of effort, and ironically produced on the Western Front the overwhelming disparity in numbers that the Allies had feared in the first place. Wary of overcommitment, deeply divided on the merits of tactical aviation, and with only a hazy intelligence picture of German intentions and strength, the Allies sheltered behind the view that whatever happened, Germany would not be strong enough to pierce the prepared battlefield. If they were less confident of defeating Germany, they were more confident of containing her, and this goal had always been the main aim of French military strategy.

The current state of air technology and military tactics clearly favored the strategic choices made by Germany in 1940 over those made by the Allies. Not only was Germany the only one of the combatants to favor a strategy of force concentration, but the campaign area itself was geographically concentrated to produce the maximum effect. During the campaign, the German air fleets massed up to 800 aircraft on any one sector of the front, while leaving other sectors entirely undefended. Or not entirely so, for the other great strength of the German advance was their use of anti-aircraft batteries, which provided very effective defense of captured positions and airfields and which the Allies had neglected as a battlefield weapon. During the course of the battle German aircraft did what they had been prepared to do: they attacked enemy air power in the field, and then concentrated the weight of attack on the rear areas and combat zone. Most of the damage was done not by dive-bombers, which constituted not much more than 10 percent of the force, were vulnerable to fighter attack, and less accurate than the popular image would suggest, but by the fast medium bombers which pounded supply columns, railways, depots, and bases in the rear. Defense of German troops and bomber formations were left to the fighter forces which operated in large groups and which quickly achieved local air superiority over the vital areas of the front.[35]

By contrast, the British and French response was famously disorganized and feeble, though perhaps less so than might have been expected from the poor operational preparation. The critical problem was the inability to produce effective force concentration. This deficiency was exacerbated by the absence of an Allied counterforce strategy, which permitted the Luftwaffe greater freedom to attack Allied air power at the source and further reduce any prospect for concentrated effort on the Allied side. As it was, the Allies never drew on more than a fraction of their available aircraft at any one time. In May 1940 only one-quarter of the combat-ready aircraft were deployed on the front facing the German attack.[36] The British sent an air contribution not much larger than the Polish Air Force that had faced the Luftwaffe in September 1939, and it got the same treatment. Its light bombers were so ineffective that within the first two days of the battle the bomber force was reduced by 50 percent.[37] The response was to send in more aircraft in small packets—a few squadrons at a time—which proved disastrous. It meant that the RAF suffered regular attrition throughout the battle, while never building up sufficient strength to stabilize its position. RAF losses were double those of the French Air Force. For the French, by contrast, the strategy of conserving forces meant that as the battle went on more reserves could be brought in, and by the end of the campaign French fighters were regaining air superiority on some parts of the front. Yet Allied air forces dissipated even this advantage by a remarkably low level of combat activity. French fighter units averaged 0.9 sorties a day; German fighters averaged 4.0. French bomber units averaged only 0.25 sorties a day.[38] This paucity can partly be explained by manpower shortages. But there were few constraints on fuel supply, or operational airfields, and the supply of new aircraft was double the German. Moreover, German airmen had problems of greater magnitude. They were often compelled to make do with makeshift airfields, with stretched lines of supply and maintenance, and the loss of pilots shot down over French territory.

Failure to concentrate air forces—at a time when the Western powers had over 1,200 fighter aircraft available—was compounded by poor communications and wildly fluctuating tactical instructions. French communications were rudimentary and preparations for coordinating air attacks in support of the army poorly carried out. The British post-campaign assessment showed that a lapse of four to five hours was standard between sighting the target, reporting back to headquarters, and dispatching an attacking force.[39] For British bombers the delay could be even longer since reports had to be sent back to headquarters in Britain first. Air-to-air and air-to-ground contact was poor by the standards of RAF Fighter Command at home, or German practice. German tank commanders could call up dive-bomber or fighter support within minutes, indeed so quickly that they were willing to send radio messages in the clear rather than in code. Target selection on the Allied side became random as the pace of the battle increased. Daylight bombing attacks were hastily withdrawn in favor of night attacks, which limited what targets could be hit effectively. Through the middle of May, the RAF chafed at the bit to be allowed to attack long-range industrial targets. Their argument now was that the bombing of German cities would reduce the pressure at the front by forcing the Germans to withdraw fighters to defend the Reich. The hope was that an enraged Hitler would also order bombers away from the battle to retaliate against Britain, where they would be destroyed on the firm shield of Fighter Command.[40] The Commanders in Chief and the politicians remained unconvinced of the virtues of such an indirect and unpredictable strategy. Only after the bombing of Rotterdam on May 14 were the gloves finally taken off and attacks on Germany permitted, but they had virtually no effect on diverting Luftwaffe aircraft from the battle. Nor would the French allow any concentrated effort against German targets. Allied bombers were divided among the battlefield, the rear areas, and distant industrial bombing. By the end of the month the two Allies were effectively fighting separate air wars. The British War Experience Committee noted in July with classic understatement: "There appears to have been some lack of touch between England and France in the latter part of May . . . ."[41] In fact, by the end of May the British had abandoned the military alliance and were intent on saving themselves. The bulk of RAF aircraft remained, as the French bitterly noted, in England.

Yet for all the deficiencies of Allied arms in May 1940, the impact of German air power should not be exaggerated. German air forces faced problems of their own and the loss rate for German aircraft was very high. In May and June 1940 the Luftwaffe lost 1,482 aircraft destroyed and 488 damaged, a total equal to almost half their total strength at the start of the battle, and well above the replacement rate from the factories.[42] As the battle continued, the lack of large air reserves began to tell and French fighter aircraft began to inflict higher losses. Allied air tactics, learned in the harsh school of battle, started to improve. Over the evacuation beaches at Dunkirk the Luftwaffe lost 240 aircraft in three days of fighting and von Kleist was forced to report "enemy air superiority."[43] Accidents, battlefield attrition, and Allied attacks on airfields all took their toll. Pilot losses were high, and by the end of the battle exhaustion and frayed nerves reduced Luftwaffe effectiveness even more. Nor was air attack as effective as it was to become later in the war. Navigation and bomb-aiming were in their infancy. Eyewitnesses noticed that it was the psychological effect of air attack that created as much havoc as actual bombing, which tended to be less accurate in battle conditions. Against a well-defended target such as Dunkirk, regular bombing even lost its psychological effect. The impact of the Battle of France was far-reaching for the Luftwaffe, reducing its aircraft numbers and skilled pilots and laying the foundation for its ultimate defeat over England a few months later. The Luftwaffe had not yet fully recovered from its mauling in France when it attacked the Soviet Union a year later. A case could be made for arguing that Allied air forces in the end did better against the Luftwaffe than might have been expected given the disparity in numbers and activity rates. The same cannot be said for the ground armies. Whatever might be presented in mitigation about Allied courage and steadfastness in the face of the enemy, the fact was that the German Army's very high standards of training and operational effectiveness, combined with the element of surprise and the concentration of armored forces in a battering ram aimed at the weakest point of the "continuous front" were all that was necessary to produce the Allied rout. Large numbers of aircraft clearly helped the ground campaign and contributed to the speed of the conquest, but the Luftwaffe failed to establish permanent air supremacy, or to pursue strategic air attacks, or to prevent the evacuation at Dunkirk. The real limits of German air power were demonstrated in the Battle of Britain in August and September 1940 when the RAF could at last fight the campaign for which it had been prepared.

Interservice rivalry, political intervention, and serious operational weaknesses contributed to reducing the effectiveness of both Allied armies and air forces. Yet the Allies' wider strategic choices were not necessarily perverse or wrong in their own terms, nor were they behind in the technical or economic race. More significant, however, for the eventual outcome of the war was the effect the campaign in the West had on the development of air power. The German armed forces remained committed to the system that had, almost fortuitously, worked so well in May 1940. Tactical aviation and combined operations remained the central role for the German Air Force for the whole of the war. Other aspects of air power—supply, air defense, air-sea cooperation, and strategic aviation—were neglected, or neglected until it was too late. In Britain the direct lessons of the campaign were slowly absorbed, and a greater tactical dimension was built into the exercise of air power when and where that became necessary. But the failure also encouraged the British to stick with what they saw as the key elements of air power, strategic bombardment and strong, independent air defenses. The United States watched the campaigns in Europe closely and concluded that tactical and strategic aviation, incorporating a counterforce capability, were interdependent components in the exercise of air power and developed its forces accordingly. If the West needed to learn any lesson from the experience of 1940, it was the indispensability of effective operational preparation. Clear operational guidelines, concentration of force, and good cooperation were there for the taking by any power.

There are also some lessons for the modern strategist. Some scarcely need to be mentioned. The establishment of a primary aim and concentration of effort to achieve it is a central principle of warfare whether defensive or offensive. Close collaboration among services and allies is always likely to achieve more than self-dependence. Both factors explain a great deal about German victory and Allied defeat in 1940. Other lessons are less clear-cut. The experience of the three warring states suggests that strategy must be matched closely to prevailing technology. This assertion is not a case for accepting obsolescence, but a case for recognizing the limitations of current technology and working within them. British air leaders between the wars embraced an optimum air strategy, and then found their technology constantly deficient. German airmen recognized air power's ultimate capabilities—and in 1939 their engineers were working on rockets, jets, and intercontinental bombers—but they also adjusted air operations to existing technical conditions. Of course, in the long run the British got to their goal, but with mixed results, and only at the expense of risking defeat in 1940 and 1941—and only, in the end, with Soviet and American assistance.

The contrast between British and German experience highlights the importance of sound intelligence in helping to form a clear strategic picture of the potential enemy. The British were so convinced that the Germans would turn their bombing weapon against them that they developed an air strategy designed to fight a very different air war from the one they actually fought in spring 1940. Western statesmen assumed that their strategy corresponded with German intentions when in practice there was a significant divergence. Armed forces can hardly be expected to fight a campaign for which they have not been adequately prepared; modern strategy has to be founded not only on the kind of war a state wants and expects to fight, but also on the kind of war the enemy actually fights. The two are not always the same. With hindsight it could be argued that RAF leaders might have responded more flexibly to the threat posed by Germany, and to the new alliance with France, though it is still important and salutary to understand why they did not.

There is one final lesson to be drawn from the Allied collapse in 1940. Military conflict is, for all its technical and organizational complexity, relatively simple to conduct. Looking back on the endless debates in the interwar years on the future nature of air power, the reader is struck by how excessively and unnecessarily complicated it had all become. It was a French air strategist, Charles Rougeron, who reminded his readers in 1938: "The essential task is fighting."[44] In 1940 German forces did not outnumber the Allies; they had no special technological lead; they were not notably more courageous; but without question the Allies were outmatched by the German ability to *fight*.

NOTES

1. A. Horne, *To Lose a Battle* (London, 1969), p. 517.

2. K. Maier, et al., *Das Deutsche Reich und der Zweite Weltkrieg: Band 2, Die Errichtung der Hegemonie auf dem europäischen Kontinent* (Stuttgart, 1979), pp. 339-40. According to the figures of the *Generalquartiermeister*, the Luftwaffe had 1,180 bombers, 341 dive-bombers, 970 single-seat fighters and 250 twin-seat fighters combat-ready on 10 May 1940.

3. F. R. Kirkland, "The French Air Force in 1940," *Air University Review* 36 (1985): 101-2; J. Truelle, "La Production aéronautique militaire française jusqu'en juin 1940," *Revue d'histoire de la Deuxième Guerre Mondiale* 19 (1969): 98-102.

4. D. Richards, *The Royal Air Force, 1939-1945, 1: The Fight at Odds* (London, 1953), 108-9.

5. De Fabribeckers, *La Campagne de l'armée Belge 1940* (Brussels, 1980), p. 36.

6. Kirkland, pp. 103, 110.

7. Public Record Office, (PRO) AIR 9/105, "British Strategical Memorandum," 20 Mar 1939, p. 41.

8. See the conclusions in M. van Creveld, *Fighting Power: German and US Army Performance 1939-1945* (London, 1983), esp. chs. 6 and 11; see too B. R. Rosen, *The Sources of Military Doctrine: France, Britain and Germany between the World Wars* (Ithaca, 1984), pp. 83-91.

9. Maier, p. 333.

10. K.-H. Volker, ed., *Dokumente und Dokumentarfotos zur Geschichte der deutschen Luftwaffe* (Stuttgart, 1968), document 200, "Luftkriegfuhrung," 1936, revised Mar 1940, p. 469.

11. E. M. Emme, "Technical Change and Western Military Thought 1914-1945," *Military Affairs* 24 (1960): 15.

12. M. Messerschmidt, "German Military Effectiveness between 1919 and 1939," in A. R. Millett, W. Murray eds., *Military Effectiveness: Vol II: The Interwar Period* (London, 1988), p. 248.

13. On strategic bombing, see R. J. Overy, "From 'Ural bomber' to 'Amerika bomber': The Luftwaffe and Strategic Bombing," *Journal of Strategic Studies* 1 (1978): 154-75; on the operational surveys, K.-H. Volker, *Die deutsche Luftwaffe 1933-1939: Führung und Rüstung der Luftwaffe* (Stuttgart, 1967), pp. 160-1, 200-01.

14. R. J. Young, "The Strategic Dream: French Air Doctrine in the Inter-War Period, 1919-1939," *Journal of Contemporary History* 9 (1974): 57-76.

15. P. Le Goyet, "Evolution de la doctrine d'emploi de l'aviation française entre 1919 et 1939," *Revue d'histoire de la Deuxième Guerre Mondiale* 19 (1969): 22-34; Général Armengaud, "L'armée de l'air et la défense nationale," *Revue des deux mondes* 103 (1933): 284-304, 545-55.

16. On US supplies for France, see J. M. Haight, *American Aid to France, 1938-1940* (New York, 1970); on army influence, P. C. F. Bankwitz, *Maxime Weygand and Civil-Military Relations in Modern France* (Cambridge, 1967), pp. 120-134.

17. R. A. Doughty, "The French Armed Forces 1918-1940," in Millett and Murray, p. 58.

18. N. N. Golovine, "Air Strategy," *Royal Air Force Quarterly* 7 (1936): pt 1: 186; pt. 3: 421. Golovine was not entirely hostile to tactical aviation, but saw it in a subsidiary light. "It is highly probable," he wrote, "that in the early stages of a future major conflict ground fighters will play an important part in cooperation with mechanized Army units." This help was included under the heading "auxiliary strategic duties."

19. PRO AIR 2/1830, *Manual of Combined Operations, 1938*, para.
20. See also M. Smith, "The RAF and Counter-force Strategy before World War II," *Journal of the Royal United Services Institution* 121 (1976): 68-72.

20. PRO AIR 2/1830, *Manual*, para. 23; AIR 9/98, "Report on Trials to Determine the Effect of Air Attack Against Aircraft Dispersed About an Aerodrome Site," Jul 1938; AIR 9/99, "Appreciation of the Employment of the British Air Striking Force against the German Air Striking Force," 26 Aug 1939, p.5.

21. PRO AIR 9/99, "Appreciation. . . ," p. 7; "Air Ministry Plans, 'Note, The Attack of Air Forces on the Ground,' 9 May 1940"; AIR 16/163, Chief of Air Staff to Bomber Command, 20 Jun 1940, p. 3.

22. PRO AIR 2/1830, Part 7.

23. PRO AIR 9/98, Bomber Command HQ, "Report on the Trials Carried Out in Connection with the Bombing of Railway Permanent Ways," Dec 1938.

24. PRO 2/7211, "Conference with the Chief of the Air Staff," 28 May 1940, pp. 1-2.

25. PRO AIR 14/181, Air Ministry "Notes on the Diving Capabilities of Monoplane Bombers," 15 Feb 1938; "Note of Bombing Error Figures, 1937/8 Dive-Bombing Trials."

26. PRO AIR 14/181, Commander, Advanced Air Striking Force to Bomber Command HQ, 5 Mar 1940.

27. PRO AIR 16/108, "Air Staff Specification A 7/39," Mar 1939; "Minutes of a Meeting of Operational Requirements Committee to Consider Army Cooperation Aeroplane," 29 Mar 1939.

28. P. C. Smith, *Impact: The Dive Bomber Pilots Speak* (London, 1981), p. 34.

29. PRO AIR 14/194, Bomber Command, "Note on the Question of Relaxing the Bombardment Instructions," 7 Sep 1939, p.6.

30. PRO AIR 14/194, General Gamelin to Air Vice-Marshal Evill, 25 Oct 1939.

31. PRO AIR 8/277, "Minutes by John Slessor," 16 Oct 1939; see also War Cabinet Conclusions, "Air Policy," Meeting of 13/14 Oct 1939.

32. PRO AIR 9/117, Anglo-French Staff Conversation, "The Attack of German Railway Communications," 26 Apr 1939.

33. PRO AIR 14/194, "Notes of a Meeting at General Gamelin's Headquarters," 24 Oct 1939, p. 5.

34. The figures are from a report sent by General Vuillemin to the French Air Minister, reproduced in *Revue d'histoire de la Deuxième Guerre Mondiale* 19 (1969): 73-4. A British report of May 1940 gave even higher figures: 4,100 German bombers and fighters with 7,000

aircraft in reserve. See PRO AIR 2/7211, "Air Ministry, Plan W.A. 1 (Modified)," 3 May 1940, p. 3.

35. Maier, pp. 340-1.

36. Kirkland, p. 102.

37. Richards, p. 118.

38. Kirkland, p. 110. French fighters nonetheless accounted for between 600 and 1,000 of all German losses.

39. PRO AIR 1/5251, "Report by the Brooke-Popham Committee," 16 Jul 1940, p. 3.

40. Richards, pp. 120-22.

41. PRO AIR 2/5251, Brooke-Popham Report, p. 3. See also P. M. H. Bell, *A Certain Eventuality: Britain and the Fall of France* (London, 1974), pp. 18-26.

42. W. Murray, *Luftwaffe: Strategy for Defeat, 1933-1945* (London, 1985), pp. 44-5.

43. Richards, p. 127. On German difficulties in preparing and conducting air operations, see K. Köhler, K.-H. Hummel, "Die Organisation der Luftwaffe, 1933-1939" in Militärgeschichtliches Forschungsamt, *Handbuch zur deutschen Militärgeschichte 7: Wehrmacht und Nationalsozialismus* (Munich, 1978): 570-79; on Luftwaffe weaknesses in 1940, see H. Faber, ed., *Luftwaffe. An Analysis by Former Luftwaffe Generals* (London, 1979), pp. 201-6.

44. "Rougeron's 'Aviation de Bombardement,' " *Royal Air Force Quarterly* 10 (1939): pt 2: 44.

PREVIOUS HARMON MEMORIAL LECTURES

I. *Why Military History*, by W. Frank Craven, 1959

II. *The Military Leadership of the North and the South*, by T. Harry Williams, 1960

III. *Pacific Command*, by Louis Morton, 1961

IV. *Operation Pointblank*, by William R. Emerson, 1962

V. *John J. Pershing and the Anatomy of Leadership*, by Frank E. Vandiver, 1963

VI. *Mr. Roosevelt's Three Wars: FDR as War Leader*, by Maurice Matloff, 1964

VII. *Problems of Coalition Warfare: The Military Alliance against Napoleon*, by Gordon A. Craig, 1965

VIII. *Innovation and Reform in Warfare*, by Peter Paret, 1966

IX. *Strategy and Policy in Twentieth-Century Warfare*, by Michael Howard, 1967

X. *George C. Marshall: Global Commander*, by Forrest C. Pogue, 1968

XI. *The War of Ideas: The United States Navy, 1870-1890*, by Etling E. Morison, 1969

XII. *The Historical Development of Contemporary Strategy*, by Theordore Ropp, 1970

XIII. *The Military in the Service of the State*, by General Sir John Hackett, G. C. B., C. B. E., D. S. O., M. C., 1971

XIV. *The Many Faces of George S. Patton, Jr.*, by Martin Blumenson, 1972

XV. *The End of Militarism*, by Russell F. Weigley, 1973

XVI. *An Enduring Challenge: The Problem of Air Force Doctrine,* by I. B. Holley, Jr., 1974

XVII. *The American Revolution Today,* by John W. Shy, 1975

XVIII. *The Young Officer in the Old Army,* by Edward M. Coffman, 1976

XIX. *The Contribution of the Frontier to the American Military Tradition,* by Robert M. Utley, 1977

XX. *The Strategist's Short Catechism: Six Questions Without Answers,* by Philip A. Crowl, 1978

XXI. *The Influence of Air Power upon Historians,* by Noel F. Parrish, 1979

XXII. *Perspectives in the History of Military Education and Professionalism,* by Richard A. Preston, 1980

XXIII. *Western Perceptions and Asian Realities,* by Akira Iriye, 1981

XXIV. *Command Crisis: MacArthur and the Korean War,* by D. Clayton James, 1982

XXV. *United Against: American Culture and Society During World War II,* by John M. Blum, 1983

XXVI. *George Washington and George Marshall: Some Reflections on the American Military Tradition,* by Don Higginbotham, 1984

XXVII. *Military Planning and National Policy: German Overtures to Two World Wars,* by Harold C. Deutsch, 1984

XXVIII. *Napoleon and Maneuver Warfare,* by Steven T. Ross, 1985

XXIX. *Soldiering in Tsarist Russia,* by John L. H. Keep, 1986

XXX. *Leadership In The Old Air Force: A Post-Graduate Assignment,* by David MacIsaac, 1987

XXXI. *The Intelligence Revolution: A Historical Perspective,* by Sir Harry Hinsley, 1988

UNITED STATES AIR FORCE ACADEMY

Provides instruction and experience to each cadet so that he or she graduates with the knowledge and character essential to leadership and with motivation to become a career officer in the United States Air Force.